BIZARRE BEAST BATTLES
ELEPHANT VS. GREAT WHITE SHARK
I0797583
Gareth Stevens
PUBLISHING
By Kate Mikoley

Please visit our website, www.garethstevens.com. For a free color catalog of all our high-quality books, call toll free 1-800-542-2595 or fax 1-877-542-2596.

Library of Congress Cataloging-in-Publication Data

Names: Mikoley, Kate, author.
Title: Elephant vs. great white shark / Kate Mikoley.
Other titles: Elephant versus great white shark
Description: New York : Gareth Stevens Publishing, [2022] | Series: Bizarre beast battles | Includes index.
Identifiers: LCCN 2020034334 (print) | LCCN 2020034335 (ebook) | ISBN 9781538264751 (library binding) | ISBN 9781538264737 (paperback) | ISBN 9781538264744 (set) | ISBN 9781538264768 (ebook)
Subjects: LCSH: Elephants—Juvenile literature. | White shark—Juvenile literature.
Classification: LCC QL737.P98 M55 2022 (print) | LCC QL737.P98 (ebook) | DDC 599.67—dc23
LC record available at https://lccn.loc.gov/2020034334
LC ebook record available at https://lccn.loc.gov/2020034335

First Edition

Published in 2022 by
Gareth Stevens Publishing
29 E. 21st Street
New York, NY 10010

Copyright © 2022 Gareth Stevens Publishing

Designer: Katelyn E. Reynolds
Editor: Therese Shea

Photo credits: Cover, p. 1 (elephant) Martyn Colbeck/Oxford Scientific/Getty Images Plus; cover, pp. 1 (shark), 13 Stephen Frink/Image Source/Getty Images Plus; cover, pp. 1–24 (background texture) Apostrophe/Shutterstock.com; pp. 4–21 (elephant icon) -ELIKA-/iStock/Getty Images Plus; pp. 4–21 (shark icon) Ace_Create/DigitalVision Vectors/Getty Images; p. 4 MrsWilkins/ iStock/Getty Images Plus; p. 5 Kenneth Canning/E+/Getty Images; p. 6 PeterHermesFurian/iStock/Getty Images Plus; p. 7 Image Source/Getty Images; pp. 8, 12, 18, 21 Manoj Shah/Stone/Getty Images; p. 9 by wildestanimal/Moment/Getty Images; p. 10 Danita Delimont/Gallo Images/Getty Images Plus; pp. 11, 21 Ken Kiefer 2/Cultura/Getty Images; p. 14 Heinrich van den Berg/The Image Bank/Getty Images Plus; p. 15 Luis Javier Sandoval Oxford Scientfic/Getty Images Plus; p. 16 Byba Sepit/Moment/Getty Images; p. 17 Alessandro De Maddalena/iStock/Getty Images Plus; p. 19 Arturo de Frias photography/Moment/Getty Images.

All rights reserved. No part of this book may be reproduced in any form without permission in writing from the publisher, except by a reviewer.

Printed in the United States of America

Some of the images in this book illustrate individuals who are models. The depictions do not imply actual situations or events.

CPSIA compliance information: Batch #CSGS22: For further information contact Gareth Stevens, New York, New York, at 1-800-542-2595.

CONTENTS

Words in the glossary appear in **bold** type the first time they are used in the text.

EXCELLENT ELEPHANTS

Elephants are known for their great size, long **trunk**, and big ears. Three species, or kinds, of elephants exist. They're each named for the areas they live in: African **savanna** elephants, African forest elephants, and Asian elephants.

Even with their big bodies, these creatures are commonly thought of as gentle giants. Still, they can **defend** themselves. Plus, elephants are strong and smart. Do you think an elephant has what it takes to battle a great white shark—and win?

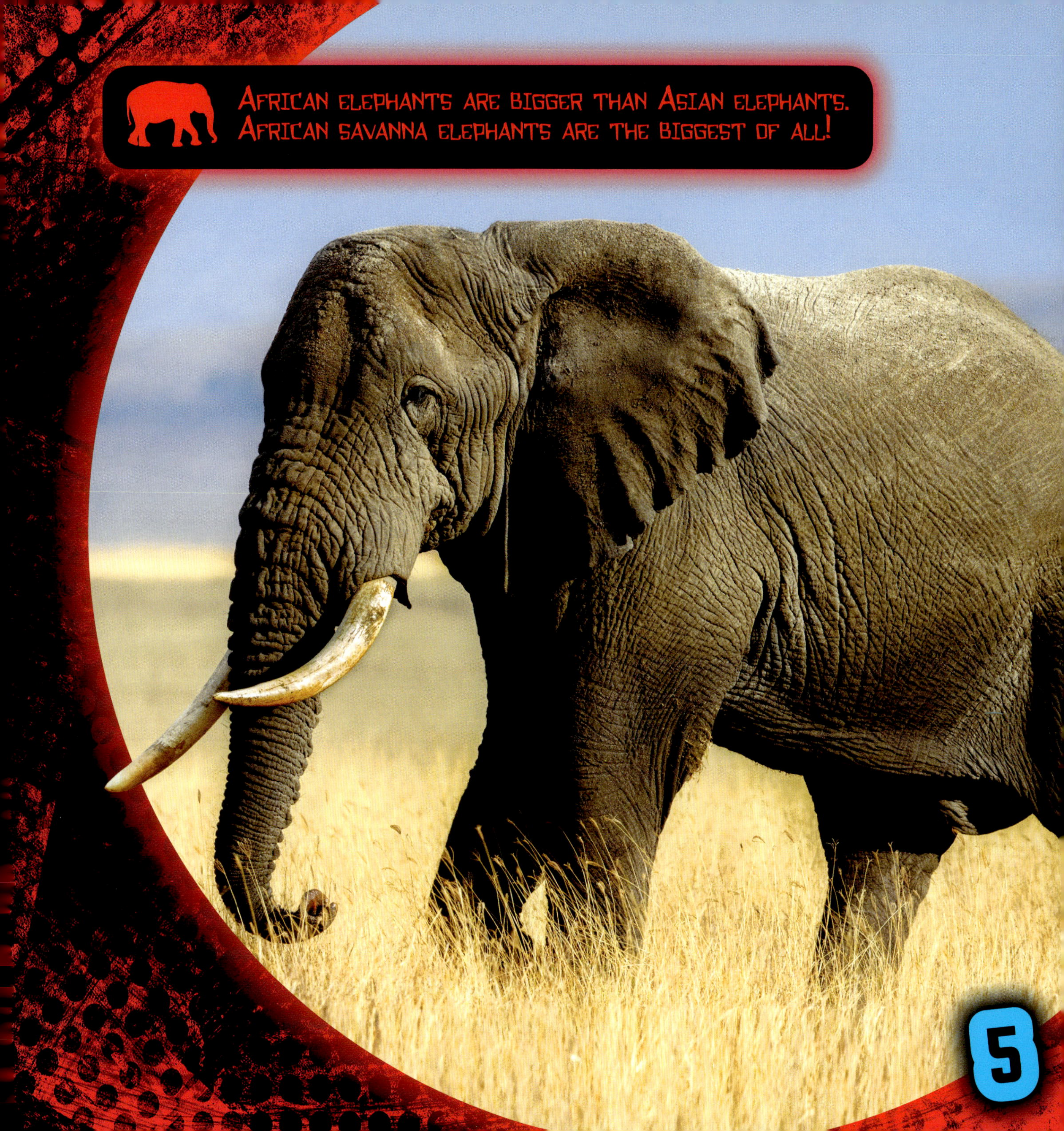

African elephants are bigger than Asian elephants. African savanna elephants are the biggest of all!

GREAT SHARKS

When you think of a great white shark, you probably think of a **fierce** swimming machine with lots of large, pointy teeth.

Are great white sharks as frightening as we think? They feast on all sorts of animals in the ocean. They rarely bite people, though. A fight between an elephant and a great white shark could be a close one. We know these sharks have the **appetite** to attack, but do they have the brains to win the battle?

GREAT WHITE SHARK RANGE

GREAT WHITE SHARKS CAN EAT A WHOLE SEAL IN JUST ONE BITE!

SIZING UP THE COMPETITION

When two animals face off in a fight, one important **factor** is size. Often, a bigger animal can overpower a smaller one. Luckily for African elephants, they're the world's largest land animals! They're up to 13 feet (4 m) tall at the shoulder.

AFRICAN ELEPHANT
LENGTH: UP TO 24 FEET (7 m)
WEIGHT: UP TO 14,000 POUNDS (6,350 KG)

GREAT WHITE SHARK

LENGTH: SOMETIMES MORE THAN 20 FEET (6 M)
WEIGHT: CAN BE MORE THAN 5,000 POUNDS (2,268 KG)

Some ocean animals are larger than elephants! Great white sharks are the largest of all predatory fish. Predatory fish are those that hunt other fish and ocean creatures. However, great white sharks are usually smaller than African elephants.

UP TO SPEED

What about speed? Elephants aren't really known for being fast. Their big bodies keep them from being able to gain much speed. Still, they can run when they have to.

ELEPHANT
TOP SPEED: 21 MILES (34 KM) PER HOUR

GREAT WHITE SHARK
TOP SPEED: 35 MILES (56 KM) PER HOUR

Great white sharks are known for being quick predators. They swim long **distances**, often at a slow pace. However, they can really pick up speed when chasing a meal. If elephants and great white sharks could race at their top speeds, the great whites would win!

TOUGH TEETH

Elephant tusks are big teeth! Only some Asian elephants don't have them. Elephants use their tusks to fight, dig, and gather food. Elephants are also born with teeth in the back of their **jaws**. As they get older, new and larger teeth grow, move forward, and **replace** the older ones. Their tusks grow too.

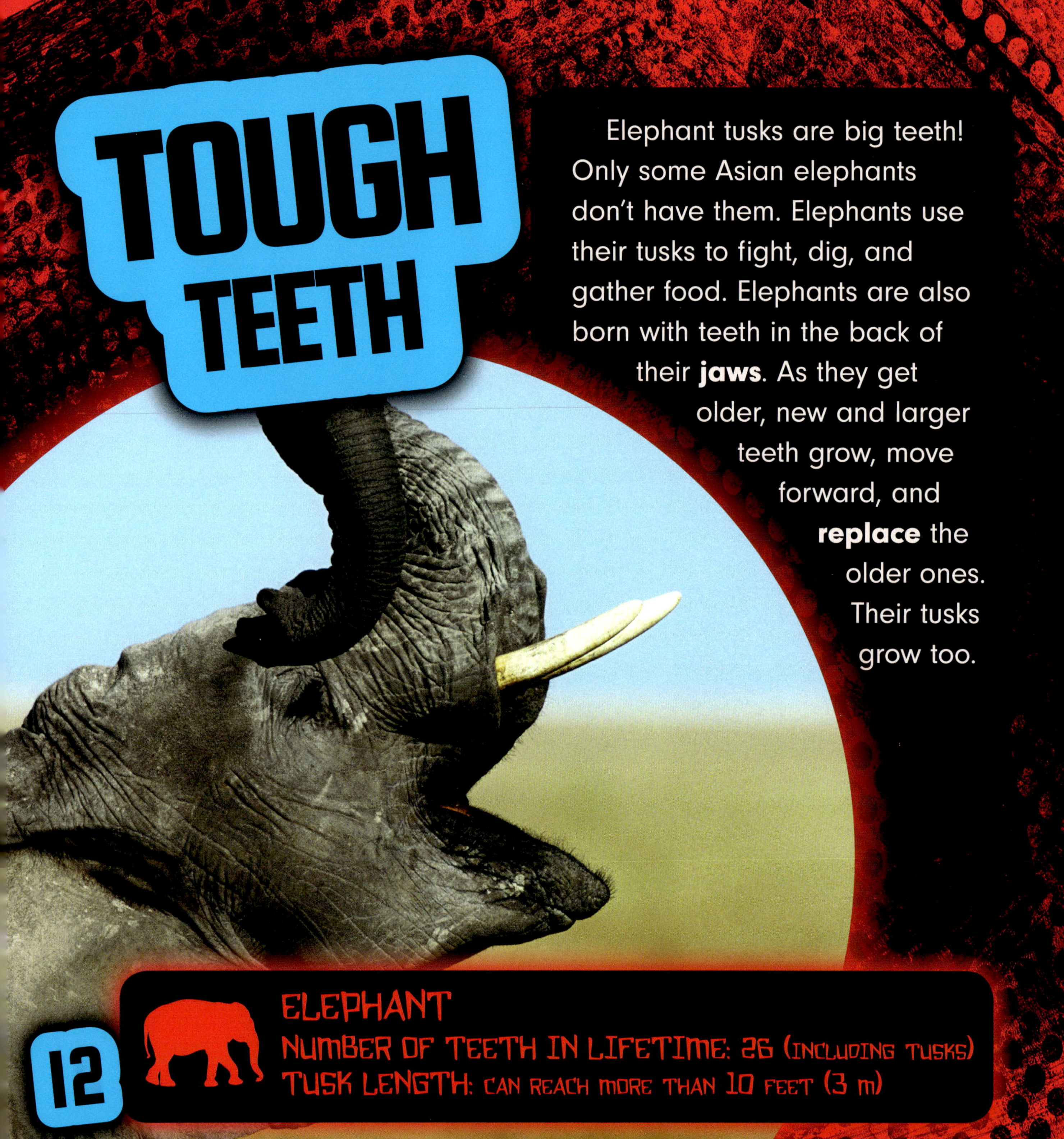

ELEPHANT

NUMBER OF TEETH IN LIFETIME: 26 (INCLUDING TUSKS)

TUSK LENGTH: CAN REACH MORE THAN 10 FEET (3 m)

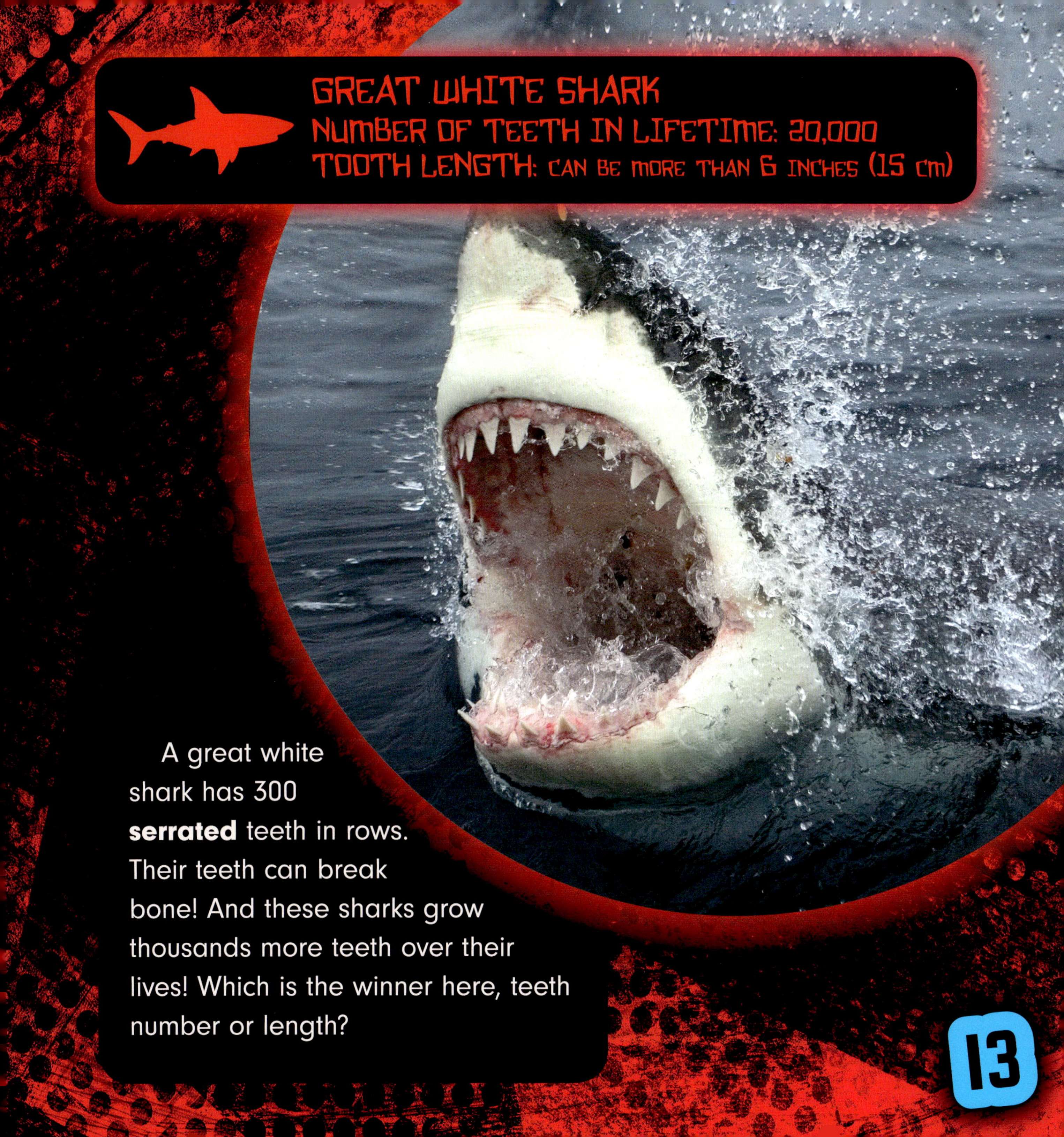

GREAT WHITE SHARK

NUMBER OF TEETH IN LIFETIME: 20,000

TOOTH LENGTH: CAN BE MORE THAN 6 INCHES (15 CM)

A great white shark has 300 **serrated** teeth in rows. Their teeth can break bone! And these sharks grow thousands more teeth over their lives! Which is the winner here, teeth number or length?

BRAINPOWER

Elephants know how to use some tools. Many animals can't. For example, an elephant in a zoo used a block as a step to reach fruit in a tree. Tools could be useful in a fight with a shark!

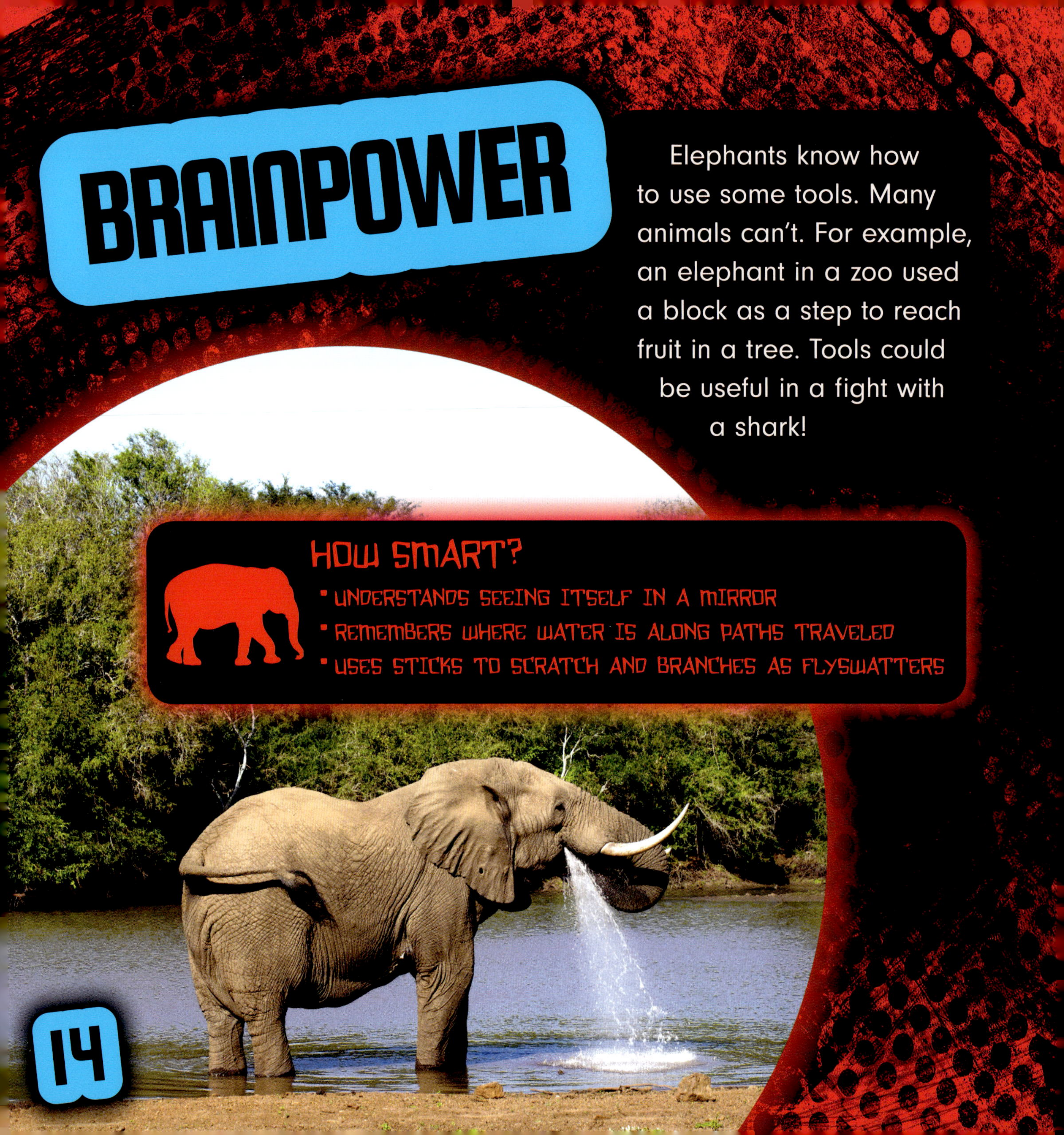

HOW SMART?

- UNDERSTANDS SEEING ITSELF IN A MIRROR
- REMEMBERS WHERE WATER IS ALONG PATHS TRAVELED
- USES STICKS TO SCRATCH AND BRANCHES AS FLYSWATTERS

HOW SMART?

- ATTACKS PREY AND WAITS NEARBY UNTIL IT DIES TO EAT IT
- CAN OUTSMART CLEVER PREY, LIKE DOLPHINS
- CURIOUS AND ABLE TO LEARN NEW THINGS QUICKLY

Great white sharks need to be smart in order to catch the animals they hunt. Great whites know to wait until just the right moment to surprise their prey. Is this contest a tie?

NEAT NOSES

An elephant's trunk is actually its upper lip and nose. But this long snout can do much more than smell! Elephants also use their trunk to eat, drink, pick up things, and **communicate**. Small hairs on the trunk are very **sensitive**.

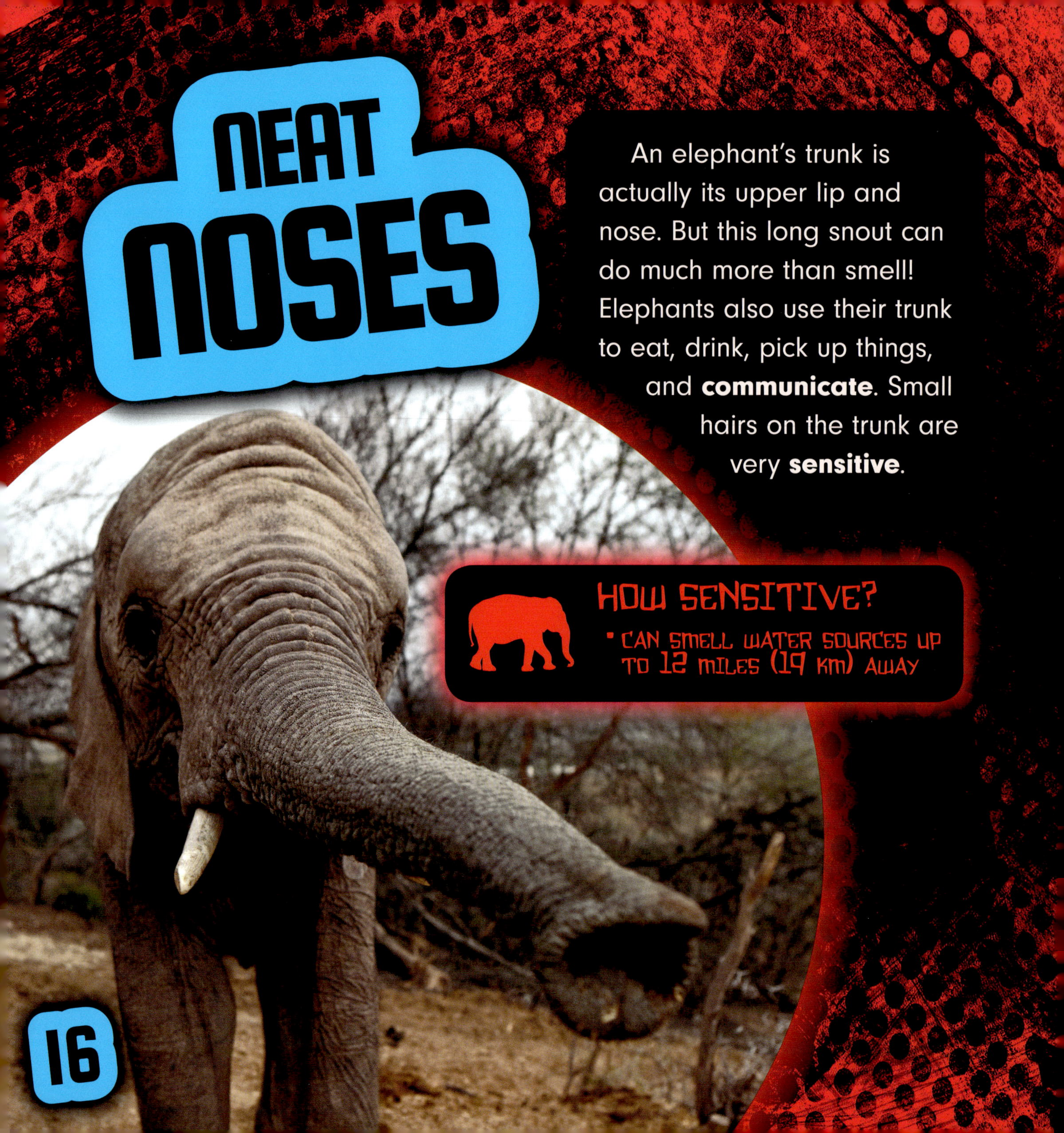

HOW SENSITIVE?

- CAN SMELL WATER SOURCES UP TO 12 MILES (19 KM) AWAY

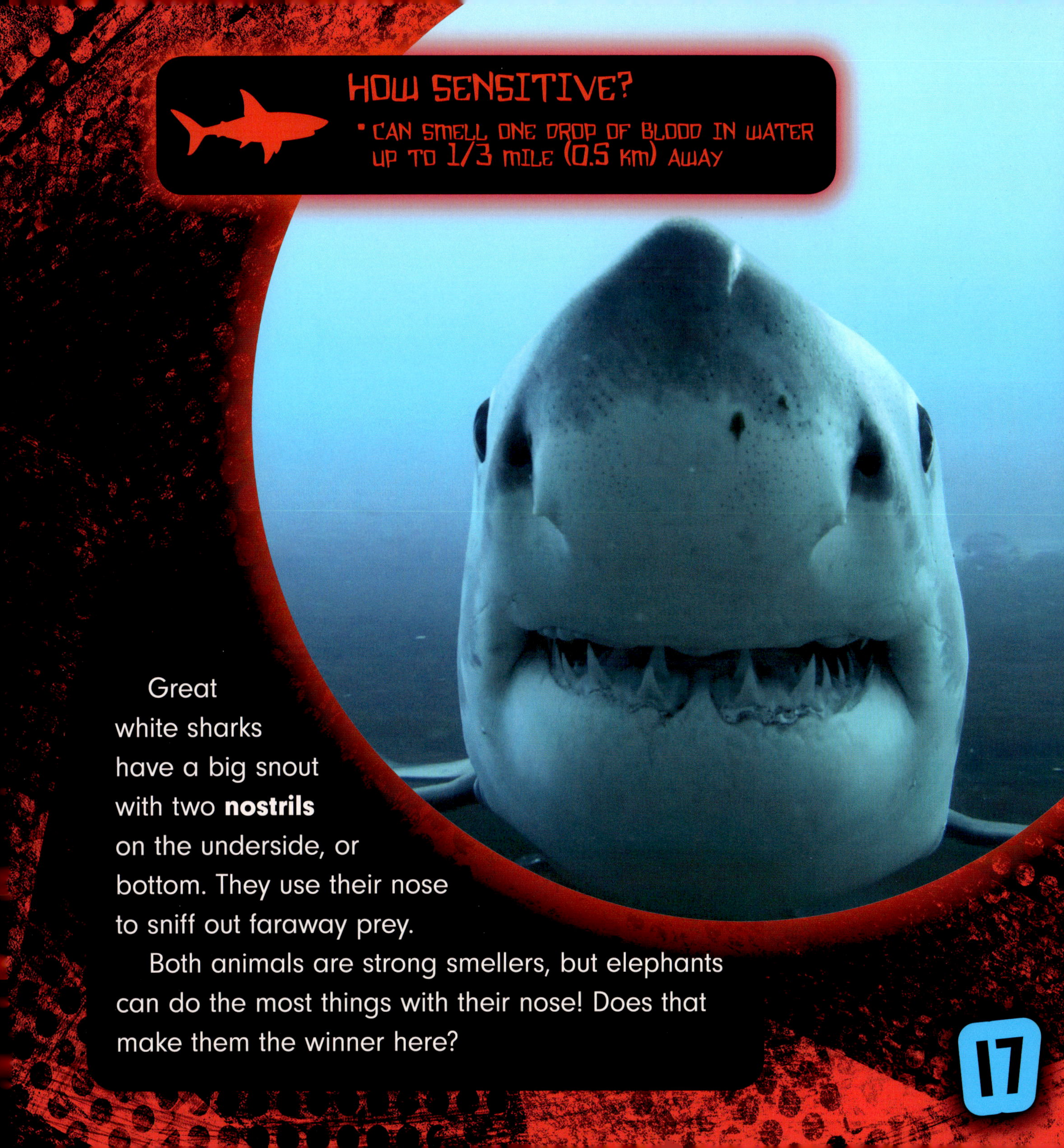

HOW SENSITIVE?

- CAN SMELL ONE DROP OF BLOOD IN WATER UP TO 1/3 MILE (0.5 KM) AWAY

Great white sharks have a big snout with two **nostrils** on the underside, or bottom. They use their nose to sniff out faraway prey.

Both animals are strong smellers, but elephants can do the most things with their nose! Does that make them the winner here?

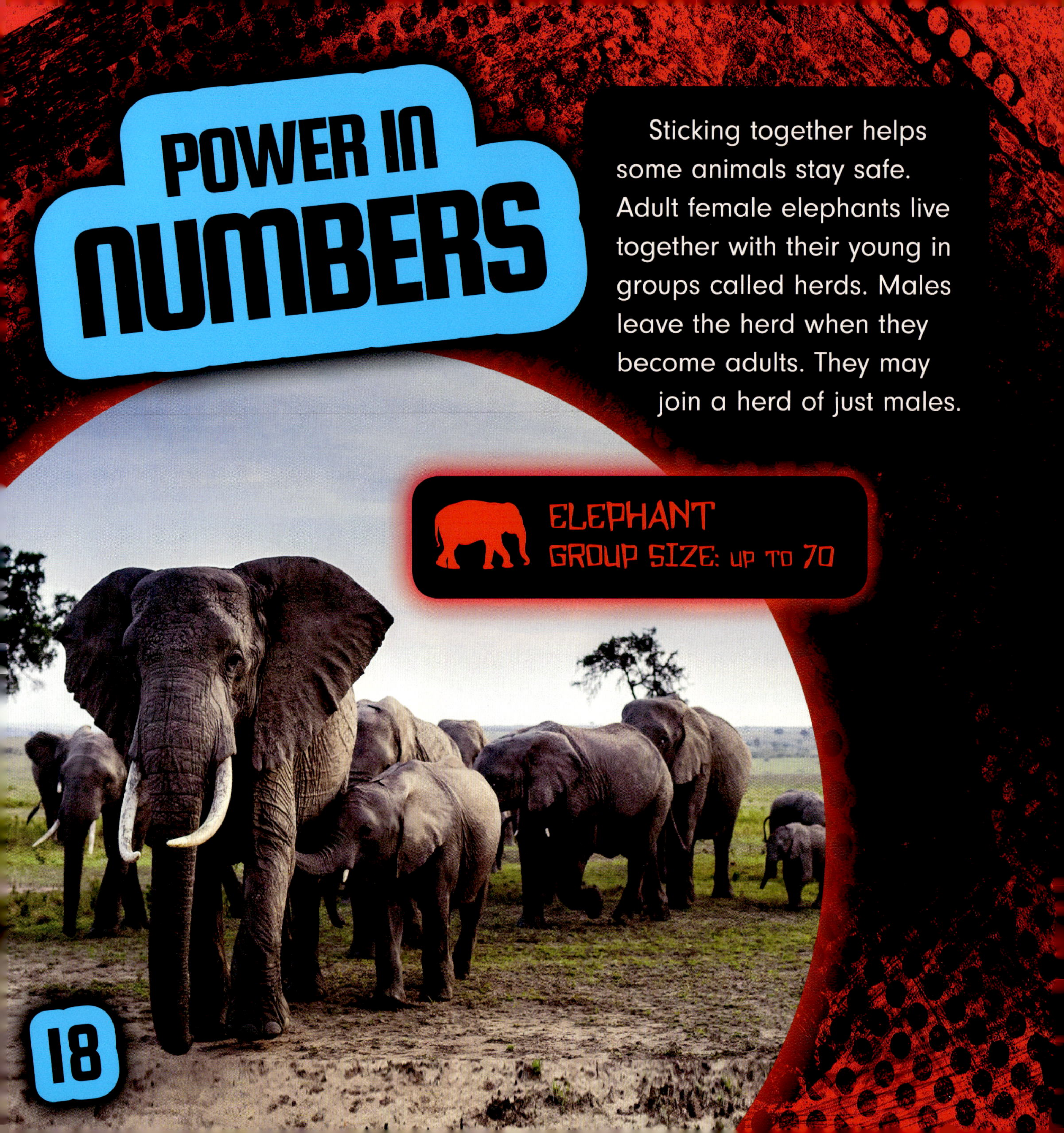

POWER IN NUMBERS

Sticking together helps some animals stay safe. Adult female elephants live together with their young in groups called herds. Males leave the herd when they become adults. They may join a herd of just males.

ELEPHANT
GROUP SIZE: UP TO 70

Great white sharks, however, swim away from their mother as soon as they're born. They commonly live alone. However, they may form a pair with another shark. Some great whites work in groups to hunt and share food. Elephants definitely have the most friends, though!

WHO WINS?

So, which do you think would win in a battle: an elephant or a great white shark? Elephants are thought of as gentle giants, while great whites are considered fierce killers. An elephant's size, trunk, tusks, brain, and herd could help it defend itself, though!

Elephants live their lives on land, while great whites survive at sea. Luckily for both animals, you aren't likely to see an elephant and a shark battle. Still, it's fun to imagine!

Both animals can live for a long time. Elephants can live to be about 60 in the wild. Great whites may live to be more than 70 years old.

GLOSSARY

appetite: an animal's want for food

communicate: to share ideas and feelings through sounds and motions

defend: to guard against harm

distance: the space between two places or things

factor: a thing that can cause something to happen

fierce: very powerful or strong, often in a harmful way

jaws: the bones that hold the teeth and make up the mouth

nostril: an opening through which an animal breathes

replace: to take the place of something else

savanna: a grassland with scattered patches of trees

sensitive: able to sense or feel changes in surroundings

serrated: having a row of small points along the side

trunk: the long nose of an elephant

BOOKS

McAneney, Caitie. *Great White Sharks*. New York, NY: PowerKids Press, 2020.

Rue, Leonard Lee. *Elephants*. Broomall, PA: Mason Crest, 2019.

WEBSITES

African Elephant
kids.nationalgeographic.com/animals/mammals/african-elephant/
Check out this page to learn more about African elephants.

Asian Elephant
kids.nationalgeographic.com/animals/mammals/asian-elephant/
Find a lot more info about these smaller elephants.

Great White Shark
www.dkfindout.com/us/animals-and-nature/fish/great-white-shark/
This website has even more fun facts about this fierce fish.

Publisher's note to educators and parents: Our editors have carefully reviewed these websites to ensure that they are suitable for students. Many websites change frequently, however, and we cannot guarantee that a site's future contents will continue to meet our high standards of quality and educational value. Be advised that students should be closely supervised whenever they access the internet.

INDEX